CUI

Fundamentals

100 Questions
(and Answers)
about the
United States
Government's
Controlled
Unclassified
Information
Program

By: James Goepel

CUI Fundamentals: 100 Questions (and Answers) About the United States Government's Controlled Unclassified Information Program

For reprint, excerpt, or other requests, the author can be emailed at Jim@FathomCyber.com or contacted via LinkedIn: https://www.linkedin.com/in/james-goepel-gc-cto-cyber/

Table of Contents

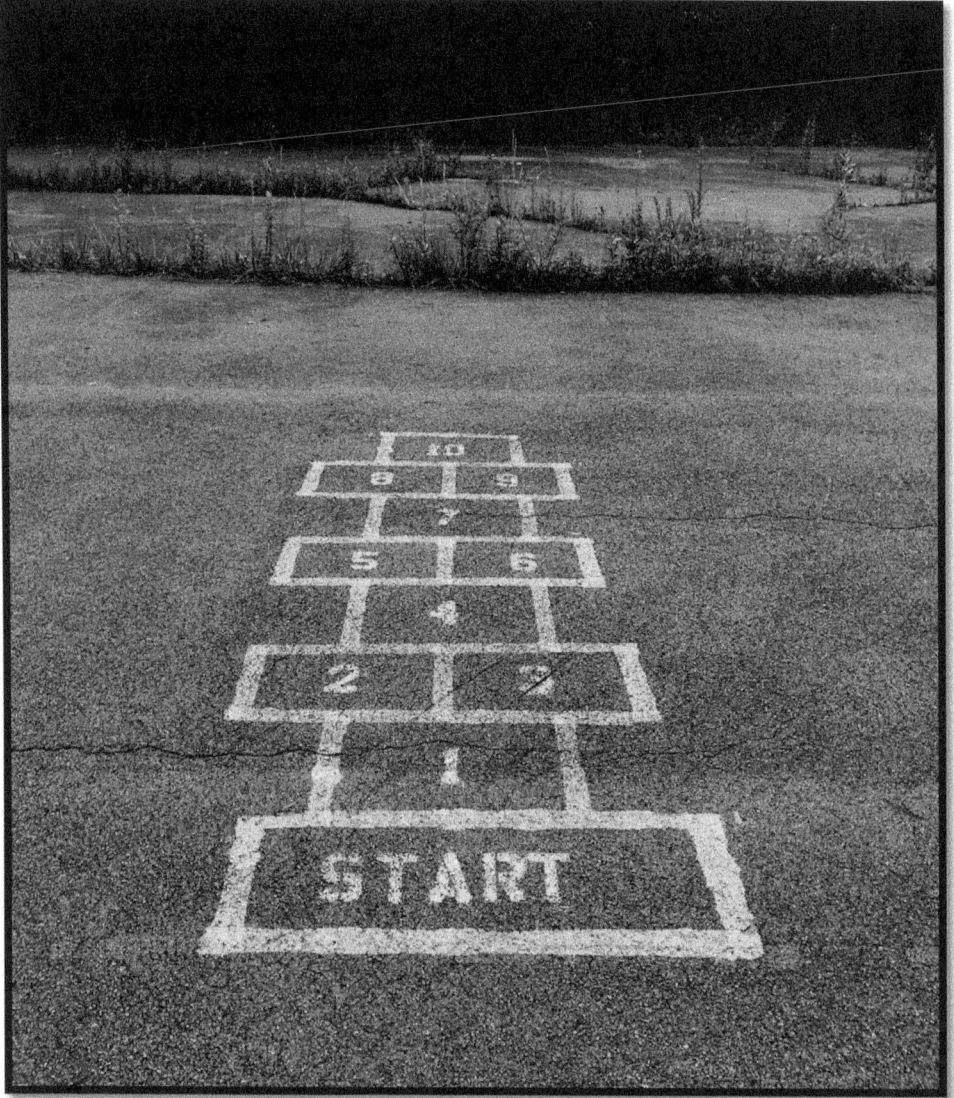

Let's get started!

Chapter 1: About This Book

1. Who is the intended audience for this book?

This book is written to help business owners, federal government employees, state and local government employees, managed service providers, and others to better understand the United States Government's Controlled Unclassified Information (CUI) program.

2. Does this book provide legal advice?

No. As with any complex topic, the details surrounding your specific use case will influence the best course of action. You should contact a lawyer with experience in government contracts and cybersecurity to discuss specifics to ensure you are meeting all applicable legal and regulatory requirements.

3. How is this book structured?

This book is written in a question-and-answer style. Each chapter focuses on a specific topic. The table of contents can help you quickly navigate to the desired topic and relevant question. However, if you are new to CUI, I encourage you to read the book from beginning to end. Many of the concepts in later chapters build on those in early chapters.

4. Why are the questions numbered?

If you bought the electronic copy of the book, there are links to certain references embedded in the book. If you bought the print copy, a list of the references is provided at the end of the book, organized by question number.

If you would like a consolidated set of electronic or hard copies of the references, they are available in the CUI Regulatory Handbook. You can find a link to the CUI Regulatory Handbook in the References section.

5. Where can I learn more about CUI?

This book is not intended as an authoritative guide on every aspect of the CUI program. If you are a lawyer, compliance professional, contracts administrator, contracting officer, or other individual who needs to know the in-and-outs of CUI and the CUI program, or if you want to learn more about the specific regulatory provisions behind a particular answer in this book, my book entitled *CUI Informed*, available from a variety of booksellers, covers the CUI program in a lot more detail. It also includes scenario-based discussions, knowledge tests, and detailed footnotes that provide additional context and support for the various topics covered, and can help you understand CUI and the CUI program at a much deeper level than this book will.

The National Archives and Records Administration (NARA) and the United States Department of Defense (DoD) also have some excellent, free training. These training programs tend to focus on CUI marking and handling, which, while critical, only cover a portion of the overall CUI program.

6. What are the differences between NARA training and DoD training?

NARA's training describes the government-wide baseline for marking and handling CUI. DoD's CUI marking training addresses DoD's implementation of the CUI program, and is mandatory for DoD staff and DoD contractors who handle CUI. This training explains how DoD expects its contractors and personnel to implement NARA's requirements. Be sure to keep a copy of your certificate once you have completed the training!

Chapter 2: CUI Program History

7. Where did the CUI program come from?

The CUI program traces its roots back to the response to the terrorist attacks of September 11, 2001 (the "9/11 attacks").

8. What motivated the government to create the CUI program?

The 9/11 attacks were a wake-up call for the federal government. Congress appointed the 9/11 Commission to study how the attackers were so easily able to avoid detection and conduct their attacks.

The 9/11 Commission found that, in fact, the attackers hadn't avoided detection. The federal government did have the information needed to identify and stop the attacks. However, due to inherent distrust among federal agencies and systemic issues with the way the government identified and shared sensitive information, the information never got to the people best positioned to take advantage of it.

The 9/11 Commission concluded that the government needed to fundamentally overhaul the way it handled and shared sensitive information, including classified and unclassified information.

9. How did the government's "need-to-know" approach to unclassified information allow for the 9/11 Attacks?

In the 9/11 Commission's view, most federal agencies had adopted a *need-to-know* approach for much of their information, including unclassified information. As the 9/11 Commission put it:

> This approach assumes it is possible to know, in advance, who will need to use the information. Such a system implicitly assumes that the risk of inadvertent disclosure outweighs the

benefits of wider sharing. Those Cold War assumptions are no longer appropriate. The culture of agencies feeling they own the information they gathered at taxpayer expense must be replaced by a culture in which the agencies instead feel they have a duty to the information—to repay the taxpayers' investment by making that information available.

Each intelligence agency has its own security practices, outgrowths of the Cold War. We certainly understand the reason for these practices. Counterintelligence concerns are still real, even if the old Soviet enemy has been replaced by other spies.

But the security concerns need to be weighed against the costs. Current security requirements nurture over-classification and excessive compartmentation of information among agencies. Each agency's incentive structure opposes sharing, with risks (criminal, civil, and internal administrative sanctions) but few rewards for sharing information. No one has to pay the long-term costs of over-classifying information, though these costs—even in literal financial terms—are substantial. There are no punishments for not sharing information. Agencies uphold a "need-to-know" culture of information protection rather than promoting a "need-to-share" culture of integration. (*The 9/11 Commission Report*, p. 417)

10. What was the 9/11 Commission's recommendation?

One of the biggest areas where the 9/11 Commission found fault was the way the government handled unclassified information. They found that different agencies categorized information using terms like *For Official Use Only* (FOUO), *Sensitive but Unclassified* (SBU), and *Law Enforcement Sensitive* (LES), without any standardization across agencies as to what qualified for inclusion in these categories. In addition, the agencies adopted similar but often conflicting rules on how various categories of sensitive information were to be protected, and with whom that information could be shared.

As a result, when, for example, Agency A sent FOUO information to Agency B, Agency B would often—at least from Agency A's perspective—"mishandle" the information. This is because, under Agency B's FOUO program, it was perfectly permissible to share FOUO with certain people inside or outside Agency B. At the same time, Agency A's FOUO program prohibited such sharing. This led to distrust between the agencies, which in turn prevented key information from reaching those who would benefit most from access to it.

NOTE: The government tends to use the terms *safeguard* rather than "protect," and *disseminate* when talking about sharing information; we'll do the same in this book going forward.

11. How did the Government respond to the 9/11 Commission's findings?

In response to the 9/11 Commission's findings, the George W. Bush White House overhauled the way the government handles classified and unclassified information. This included creating the Controlled Unclassified Information program via Executive Memo in 2008.

In 2010, the Obama White House embraced these changes and even doubled down on them by formalizing the program under Executive Order ("EO") 13556. As a result, the various federal agencies have slowly been changing their business practices to align with the CUI program and its goals, with an eye toward ensuring employees of federal agencies, state and local government employees, government contractors, international partners, and others with a lawful government purpose to do so, can access the government's sensitive, unclassified information (i.e., CUI).

12. Is the CUI program going away?

Not anytime soon. These government-wide policies have persisted through subsequent Republican and Democrat administrations, even despite complaints from various federal agencies, including those in the intelligence community.

It took a White House–appointed task force nearly six years to craft the CUI program. The CUI program has initiated a foundational change in the way the government handles unclassified information, and many agencies have been slow to embrace these changes.

Although many federal agencies have been slow to roll out their adoption of the CUI program, and some have even publicly complained about it, the White House and Congress both appear to be pushing the government to get better at sharing information to help prevent other attacks like those of September 11, 2001. Recent White House–led initiatives around cybersecurity and information security, such as the National Cybersecurity Strategy and Executive Order ("EO") 14028 on improve the nation's cybersecurity, are helping to spur these efforts on.

Chapter 3: CUI Program Overview

13. What are some of the goals of the CUI Program?

One of the goals of the CUI program is to encourage the free flow of information within—and even outside—the federal government. This is foundational not only in helping to enhance national security, but also in ensuring a healthy democracy.

Another goal of the CUI program is to ensure that sensitive information (i.e., CUI) is properly protected whenever it leaves government custody. The government has recognized that it often must entrust sensitive information to others, such as government contractors, or state and local government agencies. The CUI program defines a consistent set of safeguarding requirements for sensitive, unclassified information when it is shared outside executive branch agencies.

14. How does the CUI program achieve these goals?

The creation of the CUI program forces agencies to adopt standardized definitions for what constitutes "sensitive" information, how it must be safeguarded, and how to impose limitations on the distribution of that information.

15. Is there only one sensitivity category for unclassified information?

No. With the introduction of the CUI program, the government has created an information sensitivity spectrum. At one end of this spectrum is public information, and at the other is classified information. In the middle sits non-public, unclassified information. Although unclassified, much of this information is still valuable to cyber criminals, nation-state actors, and others who are not authorized to access it.

The government splits this unclassified, non-public information into two broad categories: Federal Contract Information (FCI) and Controlled Unclassified Information (CUI). We'll explore the differences between FCI and CUI in the next chapter but, at their most basic, FCI is all non-public information, and CUI is non-public information that is sensitive enough for Congress to have passed a law, for an agency to have passed a regulation, or for the President to have issued a government-wide policy that requires (or permits) safeguarding or dissemination controls to protect it.

16. What are some of the advantages of this new approach?

Creating these new categories and standardizing their definitions across all federal agencies will pay dividends over time. At its core, standardization allows some of the government's best and brightest cybersecurity experts to perform risk assessments based on the relative sensitivities of the different information categories. This allows the government to then create standardized approaches to safeguarding information in each of those categories.

The government, in turn, can then ensure the information can flow more freely and be made available to appropriate persons and organizations, while also ensuring that consistent safeguards are in place. It also helps build trust across agencies that previously had been reluctant to share information because of inconsistent security practices and definitions.

17. Who oversees the CUI program?

Executive Order 13556 appoints the National Archives and Records Administration (NARA) as the Executive Agent for the government-wide CUI program. NARA delegated these Executive Agent responsibilities to its Information Systems Oversight Office (ISOO), thus making the ISOO responsible for establishing and administering a consistent approach to defining and managing CUI throughout the government. The ISOO published this consistent approach as the CUI program, the requirements of which you can read in 32 CFR 2002.

18. What is the difference between the NARA CUI program and the Agency X CUI program?

32 CFR 2002 expressly cancels *legacy* markings such as FOUO and SBU, adopts a standard definition for CUI, and establishes rules for CUI designation and marking. Although 32 CFR 2002 is comprehensive, it also has some wiggle room built in that allows agencies to tailor their implementations of the CUI program to suit the way each agency conducts business.

An agency's CUI program must not conflict with any of the requirements defined in 32 CFR 2002, but the agency can add to and refine those requirements. For example, although 32 CFR 2002 identifies the unauthorized disclosure of CUI as an offense punishable by sanctions, each agency has the flexibility to define those sanctions, as well as how they are to be administered.

Chapter 4: FCI and CUI

19. Does the government care about safeguarding only CUI and classified information?

No. The government is concerned about safeguarding *all* non-public information. However, the government recognizes that safeguarding information adds to the costs borne by taxpayers and inherently restricts the free flow of that information.

Therefore, when it comes to unclassified information, the government draws a distinction between run-of-the-mill non-public information and information that is more sensitive. As discussed above, the government refers to these two broad categories of information as FCI and CUI, respectively.

20. What is FCI?

Federal Contract Information (FCI) is defined as:

> information, not intended for public release, that is provided by or generated for the Government under a contract to develop or deliver a product or service to the Government, but not including information provided by the Government to the public (such as on public websites) or simple transactional information, such as necessary to process payments. (Federal Acquisition Regulations ("FAR") 52.204-21(a))

Put more simply, FCI is information a government employee created, that someone (e.g., a government contractor) created for the government, or that someone (e.g., a citizen or a company) provided to the government, that is non-public and unclassified.

To make things a little easier to understand in this book, we'll refer to information created by the government, for the government, or given to the government, collectively as *Government Information*, even though sometimes it is technically someone else's information.

21. What are some examples of FCI?

FCI is a very broad term that includes any non-public information provided by or generated for the government under a contract. For example, a government agency might hire a marketing company to create a new advertising campaign. Until the government agency approves that campaign, the advertisements created as part of the campaign are FCI. Similarly, emails exchanged between a government contractor and a government agency may be FCI if they occur under a contract to develop or deliver a product or service to the Government.

22. Then what is CUI?

CUI is defined as:

> information the Government creates or possesses, or that an entity creates or possesses for or on behalf of the Government, that a law, regulation, or Government-wide policy requires or permits an agency to handle using safeguarding or dissemination controls. (32 CFR 2002.4(h))

23. What is the difference between CUI and FCI?

To be CUI, in addition to the information being non-public, unclassified Government Information, there must be a law, regulation, or government-wide policy (LRGWP) that makes the information CUI. If the information does not qualify for safeguarding or limited dissemination controls under a LRGWP, or if the information isn't Government Information, that information *cannot* be CUI.

24. What are some examples of CUI?

Examples of CUI include the Human Resources records of government employees, taxpayer information, patent applications, technical information with military or space application, and critical infrastructure information.

25. Is there a list of all the LRGWPs?

Yes. In its role as the CUI Executive Agent, NARA created the CUI Registry, a curated list of all the LRGWPs any agency can use as a basis for designating information as CUI. NARA updates the CUI Registry as new LRGWPs are created by the Government. Only LRGWPs in the CUI Registry may be used as the basis for designating information as CUI.

Some federal agencies, such as DoD, have published their own CUI registries. This may seem counterintuitive, since NARA's CUI Registry is the definitive source for all LRGWPs that can be the basis for CUI designation. However, these agency-specific CUI registries are subsets of the NARA CUI registry, thus narrowing the list to LRGWPs relevant or applicable to that agency.

For example, under some LRGWPs, information would only qualify as CUI when a specific agency (e.g., the Department of Agriculture) receives that information (e.g., crop yield information) as part of a particular program run by that agency. Since, in many cases, DoD is not the agency referenced in the LRGWP and is not the agency creating or receiving the information on behalf of the government, the corresponding CUI categories can be left off DoD's CUI registry. This makes it easier for those in DoD authorized to designate information as CUI, because they do not have to review as many LRGWPs as part of the designation process.

We will discuss CUI designation and marking in more detail below.

26. How many LRGWPs are in the CUI Registry?

The CUI Registry contains over 400 LRGWPs that agencies can use as the basis for designating information as CUI. Any individual piece of information may qualify as CUI under more than one LRGWP.

27. Are the LRGWPs organized in some way?

NARA has categorized the relevant LRGWPs into dozens of CUI categories based on the type of information covered. Example categories include Export Controlled, Personnel Records, General Law Enforcement, General Privacy, Death Records, Naval Nuclear Propulsion Information, General Critical Infrastructure Information, Budget, and US Census.

28. What is the difference between CUI Basic information and CUI Specified information?

Information is categorized as *CUI Specified* when the LRGWP that is the basis for designating the information as CUI also specifies or permits safeguarding or limited dissemination controls. Information is *CUI Basic* if the LRGWP does not have any explicit safeguarding or limited dissemination controls.

One example of CUI Specified is Controlled Technical Information (CTI), which you can find under the Defense Organizational Index Grouping of the CUI Registry. CTI is CUI Specified because of DFARS 252.204-7012 (the "DFARS -7012 clause").

If you aren't familiar with Defense Federal Acquisition Regulation Supplement (DFARS) contract terms, these are agency-specific that are included in contracts with DoD. These contract terms are in addition to, and supplement, those in the Federal Acquisition Regulations (FAR). FAR and DFARS clauses are often incorporated into contracts by reference. That is, the contract simply includes a reference to "DFARS 252.204-7012," without including the full language of the clause.

The DFARS -7012 clause imposes additional safeguarding requirements *and* limited dissemination controls for information with military or space applications that is created by or for DoD under a government contract.

These additional safeguarding requirements include the need to:

- use FedRAMP authorized (or equivalent) cloud providers, if DoD's CUI is placed in the cloud,
- notify DoD within certain time periods if its CUI is the subject of a data breach or other incident, and
- include limited dissemination control markings (DoD calls these *distribution statements*).

See clauses (c) through (g) of DFARS 252.204-7012 for the full list.

29. Why is the distinction between CUI Basic and CUI Specified considered "important"?

The distinction between CUI Basic and CUI Specified is important because the requirements associated with CUI Specified exist *in addition to* those associated with CUI Basic. Put more simply, if you handle information designated as CUI Specified, 32 CFR 2002 says that you must meet the safeguarding requirements defined in NIST SP 800-171 (i.e., National Institute of Standards and Technology Special Publication 800-171). Since the information is CUI Specified, 32 CFR 2002 says you must *also* meet the requirements specified in the LRGWP that is the basis for the CUI Specified designation.

Again using DoD's CTI (information with military or space applications) as an example, all CTI information must be safeguarded in accordance with the CUI Basic requirements *and* be safeguarded in accordance with the requirements in DFARS 252.204-7012(c)-(g). The fact that the DFARS -7012 clause adds additional safeguarding requirements, and limited dissemination controls, makes information designated as CTI CUI Specified.

30. Can information ever stop being CUI?

Yes. An *agency* can determine that information no longer qualifies as CUI. Only NARA and the agency that designated the information as CUI have the authority to *decontrol*, or remove, the CUI designation from that information.

31. What can cause CUI to be decontrolled?

In some cases, the agency that designates information as CUI may determine that the CUI will automatically be decontrolled at a certain future date or once a particular event occurs. This information must be communicated along with the CUI, so that anyone who receives a copy of the CUI knows when the information can be made more freely available.

Once information is decontrolled, its CUI markings should be removed.

32. Once information is decontrolled, can it be made available to the public?

In some cases, yes, but generally, no. It is important to note that decontrolling information does not automatically mean it can be released to the public, except where the LRGWP states otherwise.

In general, the decontrolled information is still FCI and must be protected from public release. In some cases, the LRGWP that made the information CUI might state that the information is to become public once it is decontrolled, but this is the exception and not the rule.

To become public, information must go through the agency's public release process. This typically involves review by authorized agency employees.

Chapter 5: Designating and Marking CUI

33. What is the difference between designating and marking information as CUI?

Designating information as CUI is the process of reviewing that information against every LRGWP in the CUI Registry to determine whether the information meets the requirements of one or more LRGWPs and thus qualifies as CUI, and then determining how the CUI should be marked. *Marking* information as CUI involves placing an appropriate CUI designation block and CUI banner marking(s) on information that has been designated as CUI. We'll explore the banner marking and designation block in more detail in question 41.

34. Who is authorized to designate information as CUI?

Designating information as CUI is an inherently governmental act not to be undertaken by merely anyone. Designating information as CUI inhibits its free flow, and the law therefore limits CUI designation to only those to whom CUI designation authority has been properly delegated.

The fact that someone is an *authorized holder* of CUI (i.e., they have a lawful government purpose to handle that CUI) does not automatically grant them authority to designate information as CUI. Individuals must have been specifically delegated authority to perform CUI designations. (See, e.g., 32 CFR 2002.20(f)(2).)

Neither government personnel nor citizens have inherent authority to designate information as CUI; doing so is fundamentally a governmental act, and authority must be officially delegated to them. These delegations are typically accomplished in conjunction with the individual's role in an agency.

To be properly delegated authority to designate information as CUI, an individual must be able to trace their authority all the way from:

- the authorization granted by the President in Executive Order 13556, to
- those delegated to the federal agencies in 32 CFR 2002, and then to
- the individual as part of their agency's CUI program.

Agencies typically delegate CUI designation authority through one or more agency memoranda or regulations that define the agency's CUI program. For example, DoD Instruction 5230.24.2.8.e includes the delegation of CUI designation authority to managers of DoD technical programs (through their ability to designate limited dissemination controls to certain information). Similarly, DoD Instruction 5200.48.3.8.d delegates to Original Classification Authorities (OCAs) the authority to designate information as CUI.

35. Does a government contractor have inherent CUI designation authority?

No. Any delegation to a contractor must be explicit in the contract and is exceedingly rare. As discussed above, under 32 CFR 2002, agencies (or more correctly, authorized employees of those agencies) are responsible for designating information as CUI. The agencies communicate CUI designation(s) to the contractor as part of the contract.

NOTE: This could be accomplished through a Security Classification Guide or as specific language in the contract, such as "The Agency has designated the XYZ software to be developed and delivered by Contractor under this contract as Controlled Unclassified Information under 12 USC 3456 and must be marked as follows: …".

36. Is delegation of authority the only issue with CUI designation?

No. Designating information as CUI also requires a careful analysis of that information against numerous LRGWPs. These LRGWPs are often written in "legalese" and can thus be difficult to interpret. Agencies must ensure that personnel are properly trained in CUI designation and marking, so that the CUI program is properly implemented in their agency. This includes training those with CUI designation authority about how to interpret the relevant LRGWPs and not merely rely on the category summaries provided as part of the CUI Registry.

37. Why do we mark information as CUI?

Sometimes the differences between CUI and FCI are subtle. For example, because of the way a particular LRGWP is written, sometimes information may not be CUI if it is created by one agency, but the exact same information might be CUI if created by another agency.

If it is CUI, the government has a legal obligation to ensure it is handled properly. The only way someone handling information can definitively know the information has been designated as CUI is if it is properly marked. If not, they may lack the proper context to determine whether it is CUI. This can allow sensitive information to be disseminated in a manner contrary to a law, regulation, or government-wide policy.

38. When must CUI be marked?

Information must be marked as CUI by the creator of that information *before* it is disseminated. Similarly, where the LRGWP applies to information that was *received* by the government (e.g., patent applications, taxpayer information, or vulnerability assessments of public drinking water systems), the government recipient must mark the information as CUI before it is disseminated, and ideally at the time it is received.

39. What is dissemination?

Dissemination is the act of "providing access, transmitting, or transferring CUI" through *any* means. CUI may only be disseminated to an authorized holder of that *particular* CUI. The disseminator must also reasonably expect the intended recipient to be capable of properly safeguarding it. We'll discuss authorized holders and reasonable expectations a little later.

40. Who determines how to mark information as CUI?

Regulation 32 CFR 2002 makes it clear that the designating agency must specify the appropriate CUI markings. Those markings must be consistent with 32 CFR 2002, but agencies do have some discretion.

41. What must be included to properly mark CUI?

CUI markings must include an appropriate CUI *banner marking*, as well as a CUI designation indicator or *designation block*, as illustrated in Figure 1 (see page 22). Where the entire document is CUI, individual portions do not need to be marked. If the information includes portions which are FCI or public information, each portion must be appropriately marked as well. Portions containing uncontrolled, unclassified information (i.e., FCI) should be marked with a (U), and CUI should be marked with (CUI). Wherever possible, CUI should be separated from the uncontrolled, unclassified information so that the uncontrolled, unclassified information can be more easily disseminated (e.g., by not including the CUI portion during dissemination to anyone who is not an authorized holder of the CUI portion).

42. What must be included in a CUI Banner marking?

A CUI banner marking includes the letters "CUI" centered at the top of the page. The banner marking also includes additional information if the

CUI is CUI Specified or if the information is subject to limited dissemination controls.

NARA'S CUI Marking Handbook has extensive information on how to properly mark many forms of CUI, including presentation slides, E-mails, documents, CD-ROMs and other removable media, portable equipment such as tablets and laptops, and much more. Rather than reprinting the CUI Marking Handbook here, please refer to the CUI Marking Handbook for more details.

43. What must be included in a CUI designation block?

A CUI designation block identifies the agency that designated the information as CUI, as well as point of contact and other information specified by the agency. This allows anyone who needs more information about the CUI, such as when and how it can be decontrolled, or the specific LRGWP(s) that make the information CUI, can know who to contact. Figure 1 includes a sample CUI designation block that meets DoD's specifications. If a contractor is creating CUI, the agency *may* authorize the contractor to list themselves as the point of contact if the contractor has rights in the information.

44. Are there any quick reference guides for marking CUI?

In addition to NARA's CUI Marking Handbook, individual agencies have created their own handbooks, trifolds, and other materials. DoD has several of these resources on their CUI page, https://www.dodcui.mil/, under Training Resources, DoD Training.

45. Where can I learn more about how to mark CUI?

NARA has a variety of free CUI-related training videos available, including videos about CUI marking. The Defense Counterintelligence and Security Agency has also published CUI marking training, which is mandatory for all DoD personnel and contractors.

If all the sub-paragraphs or sub-bullet points carry the same classification as the main paragraph or bullet point, portion marking is not required for the sub-paragraphs or sub-bullet points.

However, if any of the sub-paragraphs or sub-bullet points carry different classifications from the main paragraph or bullet point, portion marking is required for all the sub-paragraphs or sub-bullet points as demonstrated here.

Portion marks

Portions include subjects, titles, paragraphs and sub-paragraphs, bullet points and sub-bullet points, headings, pictures, graphs, charts, maps, reference list, etc.

The CUI designation indicator block does not require a portion mark.

CUI Designation Indicator

Header

Banner Marking

CUI

FOR: See Distribution

FROM: USD(I&S)

SUBJECT: (U) Information Security Considerations during Novel Coronavirus Disease (COVID-19) Mitigation Telework

(U) The President of the United States declared a National Emergency concerning the Novel Coronavirus Disease (COVID-19) outbreak on March 13, 2020. One aspect of the Federal Executive Branch's response is encouraging maximum telework flexibility. The Department of Defense is maximizing social-distancing COVID-19 mitigation efforts for all telework-ready employees.

(CUI) While the Department strongly encourages every reasonable effort to keep the DoD population and its family members and loved ones safe through social-distancing telework, we must also ensure that non-public, protected information—including Controlled Unclassified Information (CUI) and Classified National Security Information (CNSI) is safeguarded from unauthorized disclosure. Safeguarding includes a combination of physical, cyber, and other security measures.

(U) While performing COVID-19-related telework, DoD employees and contractors must make every reasonable effort to protect CUI information from unauthorized disclosure. In accordance with references (a), (c), and (d), CUI requires safeguarding measures identified in Part 2002.14 of Title 32, CFR and, as necessary, in the law, regulation, or government-wide policy with which it is associated.

(CUI) No individual may have access to CUI information unless it is determined he or she has an authorized, lawful government purpose.

(CUI) CUI information may only be shared to conduct official DoD business and must be secured from unauthorized access or exposure.

(U) Unauthorized disclosures of CUI information may result in administrative, civil, or criminal penalties, depending on the category.

Controlled by: OUSD(I&S)
Controlled by: CL&S INFOSEC
CUI Category(ies): PRVCY
Limited Dissemination Control: FEDCON
POC: John Brown, 703-555-0123

CUI

Footer

SAMPLE – NOT ACTUAL CUI

SAMPLE – NOT ACTUAL CUI

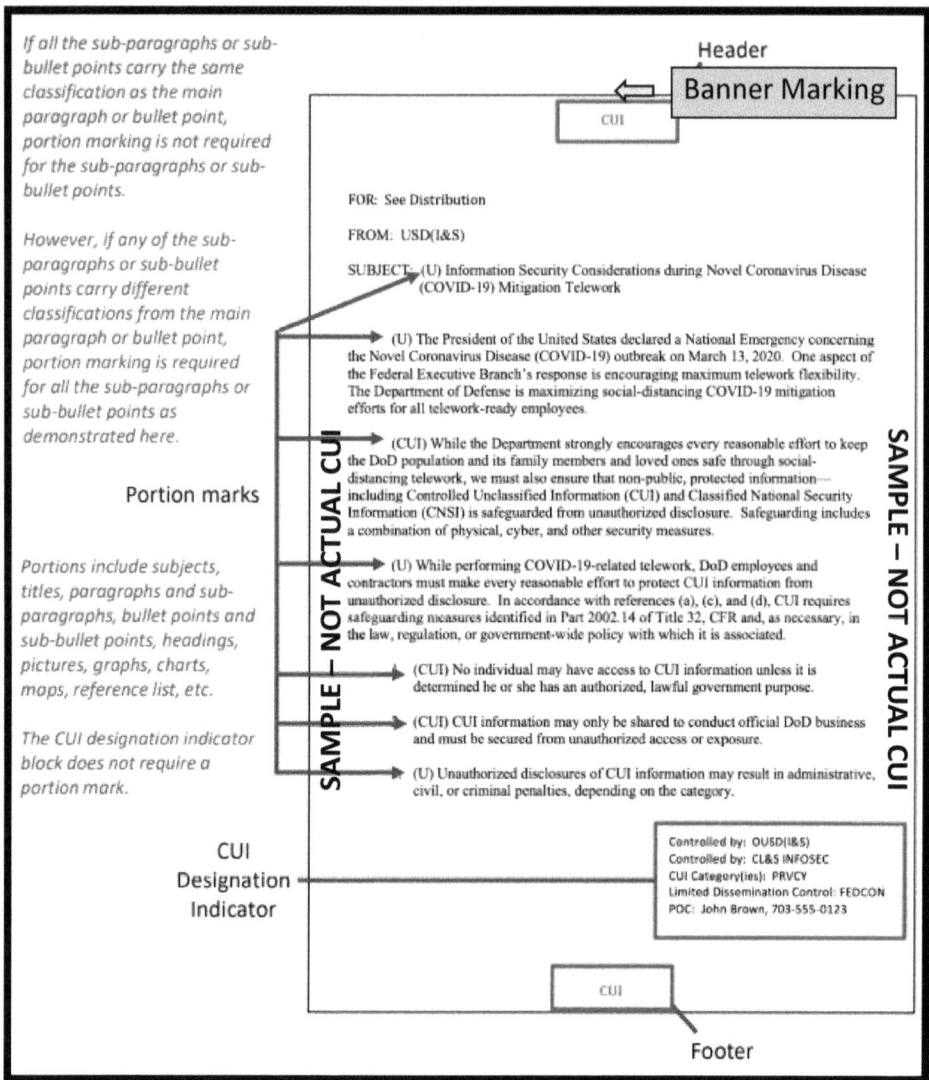

Figure 1 – *Sample* markings for CUI Basic (i.e., not actual CUI).

Chapter 6: Copies and Derivative Works

46. Must copies of CUI be marked as CUI?

Yes. A government agency has designated that information as CUI because a LRGWP requires or permits safeguarding of that information. Copying that information does not alter it, or the fact that it must be safeguarded. Therefore, all copies of CUI remain CUI and thus must be appropriately marked.

47. How should copies of CUI be marked?

Copies of CUI must be marked using the same markings as the original.

48. What is a derivative work?

Derivative works are works (e.g., drawings and excerpts) derived from, or based on, other works. For example, if DoD issues a Request for Proposals (RFP), that RFP might include a requirements specification that would likely be marked as CUI if it were to contain Controlled Technical Information. As part of the resulting contract, the contractor might create a design specification and designs derived from the requirements specification. In this case, the design specification and designs would be considered derivative works of the requirements specification.

49. Should derivative works be marked as CUI?

Most likely, yes. Continuing the example from the previous answer, the design specification and designs are derivative works of the requirements specification because they derive from the requirements specification. The design specification and designs are likely to still be CTI, since they cover the same conceptual information.

50. How do I know when my derivative work is "different enough" to no longer be CUI?

The individual creating the derivative work should have received a Security Classification Guide or other guidance from the agency when they began working on the project. The information should be compared to that guidance to determine if the information is CUI.

If a federal employee is uncertain as to whether a derivative work should be marked as CUI, they can contact the individual who designated the original information as CUI for clarification.

If a non-federal individual (e.g., a government contractor employee) is uncertain as to whether the derivative work should be marked as CUI, or believes a derivative work no longer contains CUI, the contractor should contact:

- Their prime contractor;
- The point of contact listed in the designation block in the source document;
- The point of contact listed in the SCG or contract, if one is provided;
- The tasking sponsor;
- The Technical Representative on the contract; or
- The Contracting Officer.

In approximately that order.

51. How should the derivative work be treated while awaiting an answer?

As we will discuss below, while waiting for a determination from an agency that the derivative work is/is not CUI, the individual should *treat the derivative work as though it is CUI*. If there is substantial reason to believe the derivative work no longer qualifies as CUI, the individual should *not* mark the information as CUI until it is designated as such by the agency. As a practical matter, government contractors handling suspected—but not properly designated—CUI should consider adding a cover page to the information that identifies the information as "Suspected CUI—Designation Pending" or similar markings. This will assist contractors' employees in properly handling the suspected CUI.

For DoD contractors, although DFARS 252.204-7012(m)(1) does grant defense contractors the flexibility to determine whether information, such as a derivative work, retains its nature as CUI, defense contractors are strongly encouraged to consult with the Contracting Officer, rather than make this determination on their own.

52. *Can't we just mark* all *information as CUI?*

No. Only information that qualifies for safeguarding or limited dissemination meets the definition of CUI. Misdesignating information (i.e., marking information as CUI that doesn't meet the definition of CUI) is punishable by sanctions because it can lead back to the same issues that allowed the 9/11 attacks to occur.

53. *Can we* treat *all information as though it is CUI?*

Theoretically, yes. However, implementing the safeguarding requirements associated with CUI can be significantly more expensive than simply implementing those associated with safeguarding FCI.

Rather than forcing contractors to spend more to protect information of limited value to adversaries, the government gives contractors the flexibility to handle and safeguard FCI separately from CUI. That being said, if the contractor believes it is more cost-effective or efficient to handle all government information as though it is CUI, the contractor does have that flexibility.

Chapter 7: Legacy Information

54. Is all information with legacy markings automatically CUI?

No. The 9/11 Commission found that legacy markings such as *For Official Use Only* (FOUO) and *Sensitive but Unclassified* (SBU) were often applied without any real tie back to a formal definition. This led to a lot of confusion, and unnecessary and inappropriate limitations on the sharing of that *legacy information*.

The CUI program makes those legacy markings obsolete. Regardless of the legacy marking that has been applied to any given legacy information, appropriately authorized government employees must evaluate all existing information against the relevant LRGWPs to determine if that information qualifies as CUI. It is expected that a significant portion of some agencies' legacy information will not qualify as CUI.

It is important to note that although some legacy information may not be CUI, this does *not* mean the legacy information can be made public. The legacy information must still be reviewed under the appropriate agency's public release process *prior* to any public release.

NOTE: For more information on safeguarding legacy information in government contractors' systems, see Chapter 11, question 90.

55. When must information with legacy markings be reviewed?

Information with legacy markings must be reviewed against the relevant LRGWPs *before* it is disseminated outside a federal agency. The legacy markings must likewise be removed from the information before it is disseminated and, if appropriate, CUI markings must be added.

56. What can happen to someone who misdesignates information as CUI or otherwise mishandles CUI?

Regulation 32 CFR 2002 takes seriously the designation and marking of information as CUI, and authorizes the imposition of sanctions on anyone who misdesignates information as CUI. In this context, misdesignation includes intentionally putting CUI markings on information that has *not* been designated as CUI by an agency, as well as failing to mark information that *has* been designated as CUI. The regulation also authorizes the imposition of sanctions on anyone who mishandles information that *is* CUI. In this context, mishandling includes disseminating CUI (including providing access) to someone who is not an authorized holder.

Given the possibility of sanctions, government staff and contractors should be very careful to properly identify and protect CUI. This includes ensuring that CUI is only disseminated to authorized holders, and that CUI is properly marked prior to dissemination.

Government employees must also resist the urge to simply slap CUI markings on legacy or other information, or instruct contractors to do so. Instead, an appropriately authorized agency representative must carefully review that information and designate it as CUI, if deemed appropriate.

Government contractors must not place CUI markings on legacy information. They must wait for confirmation and information from the appropriate agency (i.e., the agency whose information is in question).

Chapter 8: Accessing and Disseminating CUI

57. Can I disseminate CUI publicly?

No. Recall that CUI is, by definition, nonpublic information. So that information must not be disseminated publicly. Although the CUI program encourages the sharing of information, 32 CFR 2002 states that anyone authorized to handle CUI "should disseminate and encourage access to CUI Basic for any recipient *when the access meets [certain requirements]*" (emphasis added). The general public does not meet the necessary criteria for access, and therefore should be prohibited from accessing the CUI.

CUI Specified information has even tighter dissemination controls. It should only be disseminated, and access should only be permitted "as required or permitted by the authorizing laws, regulations, or Government-wide policies that established that CUI Specified."

Next, we'll walk through these concepts in greater detail.

58. To whom can CUI Basic be disseminated?

32 CFR 2002 states that:

> Agencies should disseminate and permit access to CUI, provided such assess or dissemination ... furthers a lawful government purpose. (32 CFR 2002.16(a)(1))

This means CUI is meant to be shared (i.e., disseminated or access is granted to it), provided such sharing is consistent with a lawful government purpose.

59. What is a lawful government purpose?

32 CFR 2002 defines lawful government purpose as:

> any activity, mission, function, operation, or endeavor that the U.S. Government authorizes or recognizes as within the scope of its legal authorities or the legal authorities of non-executive branch entities (such as state and local law enforcement). (32 CFR 2002.4(bb))

That definition gives lawful government purpose a very broad scope and could include IT service providers (who need access to the CUI to administer the systems on which the CUI is stored), as well as contractors and subcontractors. It can also include personnel at other federal agencies who, due to their roles and responsibilities at those agencies, would benefit from having access to the CUI. But it would most likely *not* include janitorial or cafeteria staff because, given their roles, they would not be likely to have a lawful government purpose to access the information.

60. What is the difference between lawful government purpose and need-to-know?

Lawful government purpose has a much broader scope than *need-to-know*. Executive order 13526, which establishes the classified information program, includes the following definition:

> "Need-to-know" means a determination within the executive branch in accordance with directives issued pursuant to this order that a prospective recipient requires access to specific classified information in order to perform or assist in a lawful and authorized governmental function. (EO 13526 Sec. 6.1(dd))

61. That didn't help; can you put it in plain English?

Need-to-know means that the recipient must meet three critical requirements to be granted access:

- approval by authorized executive branch personnel,
- *requirement* to further a government function, and
- ties to a lawful *and* authorized government function.

Needless to say, these requirements are a big hurdle to overcome!

By contrast, *lawful government* purpose means that the recipient has a valid reason, recognized by the US government and within the scope of the law, for access to the information. This means, for example, that it would likely be acceptable for a prime contractor to determine that it would be beneficial to give a subcontractor access to a specific item of CUI to help push a contract forward.

Similarly, a government employee could realize that the information they hold would be beneficial to someone in another agency and thus proactively share it with them. That is a significant shift from the need-to-know requirements for classified information, which requires approval by an Original Classification Authority prior to information dissemination.

It is important to note that, although lawful government purpose is a much more permissive approach to information dissemination than is typically applied in the classified information program, as discussed above, there are still many scenarios in which individuals are not authorized holders because they do not have a lawful government purpose to access that information.

62. Are there any other limitations on disseminating CUI?

Yes. Regardless of whether it is CUI Basic or CUI Specified, CUI can only be disseminated to authorized holders. To be an *authorized holder*, the intended recipient must, at a minimum:

a) have a lawful government purpose to access or receive that CUI,

b) understand how to properly safeguard the CUI, and

c) not be precluded from accessing or receiving the CUI due to limited dissemination control.

We'll walk through each of these requirements in a little more detail, below, beginning with question 65.

63. Are there any additional dissemination limitations on CUI Specified?

Yes. In addition to the limitations on CUI Basic, you also must ensure that the dissemination does not conflict with the specific requirements in the LRGWP that forms the basis for the CUI designation.

64. If someone is an authorized holder of one type of CUI, are they an authorized holder for all types of CUI?

No. The analysis must be performed on an information-by-information basis. For example, although Mary may have a lawful government purpose to access social security numbers because she works in the Department of Good Works' (DGW) Human Resources Department, she probably does not have a lawful government purpose to access Covered Defense Information shared by DoD with DGW.

65. What other limitations are there when disseminating CUI?

In addition to determining that the intended recipient has a lawful government purpose to access or receive the information you wish to disseminate/permit access to, you must also:

- have a reasonable expectation that the recipient understands how to handle the information, and

- ensure that the recipient is not prohibited from accessing the information due to a limited dissemination control.

66. What does it mean to have "reasonable expectations"?

It means you can't just disseminate or grant someone access to the CUI. You must ask questions—ideally *pointed* questions—of the intended recipient before you give them the CUI.

Be careful as you implement your "reasonable expectations" program. The intended recipient's mere response to the questionnaire is *not* sufficient to establish reasonable expectations. You also must thoroughly *review the intended recipient's responses* to verify and establish confidence in the intended recipient's ability to handle the CUI.

Ideally, you should document the results of your review to help prove the review was conducted.

67. What kinds of questions should be asked?

The CMMC Information Institute (https://CMMCInfo.org) has published a free sample questionnaire, which you can use as a basis for determining whether the intended recipient is capable of handling CUI.

68. What are limited dissemination controls?

Limited dissemination controls, as their name suggests, limit to whom CUI can be distributed. Agencies may place limits on disseminating CUI (beyond "for a lawful Government purpose") only through the use of the limited dissemination controls listed in the CUI Registry, and through methods authorized by a CUI Specified LRGWP.

69. What are some examples of limited dissemination controls?

Some of the more common, authorized, limited dissemination controls include those listed in the table below.

Limited Dissemination Control	Description	Marking
No foreign dissemination	Information may not be disseminated in any form to foreign governments, foreign nationals, foreign or international organizations, or non-US citizens.	NOFORN
Federal employees only	Dissemination authorized only to (1) employees of United States Government executive branch departments and agencies (as the agency is defined in 5 U.S.C. 105), or (2) armed forces personnel of the United States or Active Guard and Reserve (as defined in 10 USC 101).	FED ONLY
Federal employees and contractors only	Dissemination authorized only to (1) employees of United States Government executive branch departments and agencies (as the agency is defined in 5 U.S.C. 105), (2) armed forces personnel of the United States or Active Guard and Reserve (as defined in 10 USC 101), or (3) individuals or employers who enter into a contract with the United States (any department or agency) to perform a specific job, supply labor and materials, or for the sale of products and services, so long as dissemination is in furtherance of that contractual purpose.	FEDCON
No dissemination to contractors	No dissemination authorized to individuals or employers who enter into a contract with the United States (any department or agency) to perform a specific job, supply labor and materials, or for the sale of products and services. Note: This dissemination control is intended for use when dissemination is not permitted to Federal contractors, but permits dissemination to state, local, or tribal employees. No dissemination authorized to individuals or employers who enter into a contract with the United States (any department or agency) to perform a specific job, supply labor and materials, or for the sale of products and services.	NOCON

70. Can agencies create their own limited dissemination controls?

Yes and no. Agencies *must* use the limited dissemination controls in the CUI registry, unless they go through the regulatory process and create new limited dissemination controls as part of a LRGWP. For example, DoD created DFARS 252.204-7012, which allows marking of CUI using the limited dissemination controls specified in DoD Instruction 5230.24.

71. What should I do if I accidentally disseminate CUI to someone who isn't an authorized holder?

It depends. If you are a federal employee, follow your agency's notification requirements. These should be specified in a regulation, policy memo, instruction, or similar document. If your agency has not established notification requirements, notify the disseminating agency (i.e., the one that gave it to you) or the designating agency (i.e., the one that designated the information as CUI), if you are not certain of the disseminating agency. NARA maintains a list of CUI points of contact for most agencies, which you can use to contact the designating agency. If all else fails, you can contact the CUI Executive Agent at the ISOO.

If you are a federal contractor, you should be sure your incident response plan addresses "spills" (i.e., unauthorized disclosures). The incident response plan should include instructions for employees to notify your organization's CUI program manager or security point of contact, if you have one.

From there, either the employee or the organization's CUI program manager would contact the prime contractor, if there is one. The spill would then be escalated to the Contracting Officer or other point of contact identified in the contract. That should be done by your organization if you are the prime contractor, or the prime contractor if there is one.

If the Contracting Officer cannot be reached or is not sure how to move forward, the spill can be reported to the agency's CUI point of contact. If all else fails, the spill can be reported to the CUI Executive Agent (i.e. the ISOO at NARA).

72. Does accidental dissemination of CUI automatically make it public information?

No. The information remains CUI until the designating agency or NARA decontrol the information.

Chapter 9: Receiving CUI

73. If information isn't marked as CUI, then it isn't CUI, right?

No. CUI is *supposed* to be properly marked before it is disseminated.

However, recall that, while a goal of the CUI program is to encourage the free flow of information, another is to ensure that CUI is properly safeguarded. This includes scenarios in which people make mistakes (we're all human, after all) and forget to mark information as CUI.

In addition, although there is a strong preference for marking the information itself, information can be "marked" by contract. That is, a contract with the government can state that certain kinds of information the government provides to a third party (e.g., another agency or government contractor) is CUI. Thus, the information may have been designated and "marked" as CUI without any marking appearing on the information.

Yes, that makes things confusing! It is also one of the many reasons why it is important to carefully read and understand *every* part of any contract, including (and especially) the clauses incorporated by reference.

74. What should I do with information I received that I think is CUI but isn't marked as CUI?

First off, don't panic.

Unusual use case

If you are working for the government and receive information from someone outside the government that your agency has designated as CUI (e.g., taxpayer information if you work for the IRS, patent applications if you work for the USPTO), then since the information has previously been designated as CUI, you *must mark* the information as CUI and handle it accordingly.

Other cases

In all other cases, begin by treating, but *not* marking, the information as CUI. That is, limit access only to those with a lawful government purpose, store it in a safe place, and so on.

The reason you shouldn't mark it as CUI is that you are not the creator of that information. The creator's agency is accountable for ensuring that the CUI created in their agency has been properly designated, and the creator is solely accountable for ensuring CUI they create is properly marked. You, as the recipient of the information, are generally not authorized to designate CUI.

Next, contact the information creator or agency that disseminated the information to you. Get their feedback on whether the information was accidentally disseminated to you without CUI markings. If it was, then destroy the copy you received and ask them to send you a properly marked copy.

Doing so helps avoid several potentially impactful scenarios, such as where information was publicly released (making it neither CUI nor FCI), but a recipient (e.g., a third-tier subcontractor) is unaware that it is public information and believes it should be CUI. If that recipient marks the information as CUI and sends it back to others on the contract (e.g., their mid-tier and prime contractor), it can set off a chain of events that could take weeks—if not months—to unravel, resulting in significant inefficiencies and added expense under the contract.

75. What should I do with information marked as CUI Specified, but which does not include the LRGWP to look to for the additional requirements?

Recall that information is CUI Specified if a LRGWP says it must, or can, be subject to additional safeguarding or dissemination controls. Ideally, those additional safeguarding or dissemination controls should be conveyed to you by the authorized holder who disseminated the information, such as in a memo, in an agreement (e.g., contract) with you, in a Security Classification Guide, or in the designation block on the CUI itself. As a practical matter, since most agencies are still learning their way around the CUI program, this may not always happen.

In some cases, it is easy to identify the relevant additional safeguarding or dissemination controls, because the CUI category (indicated by the CUI marking) only includes a single CUI Specified LRGWP. In this case, you should review the corresponding LRGWP to determine the corresponding requirements and apply them accordingly.

In several other cases, the CUI categories include numerous CUI Specified LRGWPs. This can make it difficult to identify which one is the basis for the CUI Specified designation. In this case, your best bet is to contact the point of contact in the source document, contractor, or the security classification guide. If they are unresponsive or are not the designating agency, you may need to contact the designating agency and that agency's CUI point of contact.

Again, designating agencies can avoid this conundrum (and being inundated with calls/emails) by including a reference to the LRGWP in the designation block.

Chapter 10: Safeguarding the Government's Information

76. Must FCI be protected?

Yes. Even though it is not classified and is not covered by a LRGWP, FCI must still be protected. Regulation FAR 52.204-21 defines the minimum protections that must be in place when FCI is in government contractors' IT systems.

77. How must CUI be safeguarded in federal IT systems?

If CUI is in the federal government's IT systems, 32 CFR 2002 says that the agency handling the CUI, or contractors acting on behalf of the agency, must categorize CUI Basic at a moderate (or higher) confidentiality impact level under FIPS-199. Agencies must then apply the security controls described in FIPS-200 and NIST SP 800-53 in accordance with any risk-based tailoring decisions made by the agency.

If that all sounds like a big bundle of confusion to you, you aren't alone. Thankfully, the government is in the process of creating a standardized set of requirements for safeguarding unclassified information, including CUI, in government systems. This should further help build trust between agencies and allow for better information sharing. An updated version of this book will be available shortly after that regulation is available and effective.

78. Do those same requirements apply to non-federal systems?

Currently, no. If the CUI is in a non-federal system, such as a government contractor's IT system, that system must fully implement the requirements defined in NIST SP 800-171.

In the previous answer, we discussed that the government is coming out with a standardized set of requirements for protecting CUI in federal systems. There are suspicions that those requirements will closely align with NIST SP 800-171. So, at some point, federal and non-federal systems will likely have very similar, if not identical, protection requirements.

79. Is NIST SP 800-171 a cybersecurity framework?

Yes and no. Remember that the CUI program was created to protect all CUI, including electronic and physical copies of CUI. As you might expect, when NARA worked with NIST to create NIST SP 800-171, they ensured it was designed to protect CUI Basic in its various forms.

Thus, while NIST SP 800-171 includes cybersecurity-related requirements, NIST SP 800-171's requirements are not limited to cybersecurity. That is, NIST SP 800-171, and the CUI program in general, is concerned with more than just the protection of electronic copies of CUI as it moves around your IT system or the Internet. The CUI program and NIST SP 800-171 are also concerned with physical security of the computer equipment that handles the CUI, the security of paper copies of CUI, and even copies of CUI that exist on removable media such as USB drives and CD-ROMs.

80. How do we know if our company meets the requirements of NIST SP 800-171?

NIST published NIST SP 800-171A, which is the *assessment guide* for NIST SP 800-171. The assessment guide expands the 110 requirements defined in NIST SP 800-171 to a total of 320 *objectives*, or outcomes, that your information security program must meet.

To determine whether your organization complies with NIST SP 800-171, compare your organization's information security program against the objectives in NIST SP 800-171A. If the organization meets all the applicable objectives associated with a given requirement, then your organization meets that requirement. Until all applicable objectives associated with all requirements are met, your organization is not in compliance with NIST SP 800-171.

81. What makes a requirement or objective "not applicable"?

The short answer is that all objectives are likely to be applicable. The only time they are *not applicable* is if there is no relevant component of your organization's information security program.

For example, some of the requirements in NIST SP 800-171 deal with securing wireless access (e.g., Wi-Fi) to your system. If the system does not support or permit Wi-Fi or other wireless access, then those requirements would not be applicable. Some agencies require agency approval before a requirement can be deemed not applicable.

82. Do we need policies and procedures for every requirement?

Not necessarily. Under NIST SP 800-171 Rev. 2 (the current revision of NIST SP 800-171), some requirements could be implemented in certain environments in such a way that a formal, written policy or procedure is not necessary. In other cases, other types of documentation, such as diagrams or specifications, can be used in place of a policy or procedure.

However, from an evidentiary point of view, it just makes things easier to have written policies, procedures, plans, and so on, that define the organization's information security program. This makes auditing and assessments easier, and makes it easier for everyone in the organization, because the policies exist in writing and in one central, readily referenced location.

83. What is the difference between a policy, a procedure, and a plan?

A *policy* is a statement by the organization that establishes overarching guidance on a particular issue. A *procedure* is a checklist-style process that is followed to implement the policy in a given context. A *plan* is like a procedure, except that decisions need to be made during the execution of the procedure. Thus, unlike a procedure, which produces the same outcome every time it is followed, following a plan several times may result in different outcomes due to decisions made.

In general, procedures are written to be followed by lower-level, or more junior, employees who do not have decision-making authority for the organization. Plans are often executed by individuals at the management level who have at least limited decision-making authority.

84. Do we need evidence of how we meet a requirement?

Yes.

85. What kind(s) of evidence should we collect?

The goal is to show that your organization has adopted all the requirements, and that the corresponding policies, procedures, and plans are followed on an ongoing basis. The only way to show this is by collecting appropriate evidence on a periodic or per-occurrence basis.

Under NIST SP 800-171, your organization should focus on defining three categories of evidence:

- Examine
- Interview
- Test

  Examine evidence includes policies, procedures, plans, drawings, and other documentation that helps establish the way the company intends to operate. Many documents, such as an information security policy, often explicitly address multiple requirements, which is perfectly acceptable.

Interview evidence is a list of individuals accountable and/or responsible for ensuring that a particular requirement is implemented by the organization. As this term implies, this is collected to allow auditors and assessors to interview the relevant personnel and confirm that their understanding of what the organization is doing is consistent with both the information in the Examine documents and the requirements in NIST SP 800-171.

Test evidence involves the auditor or assessor viewing the organization in operation to confirm that actual business practices align with the Examine and Interview evidence. To expedite the testing process, it is often beneficial to collect a list of relevant tools or services used as part of the system being audited or assessed.

86. Can our cleaning crew access rooms where CUI is stored?

It depends. Recall that dissemination can include *any* kind of access, including reading CUI over someone's shoulders or from their desk after hours. The cleaning crew most likely does not have a lawful government purpose for accessing the CUI. Therefore, they must not be permitted to enter the room if the CUI is out in the open or is otherwise accessible.

If all electronic copies of the information are stored in locked computer systems, and paper copies are in locked filing cabinets (and the cleaning crew cannot easily unlock them), then, assuming the organization has a "clean desk" policy that requires that all CUI be put away every night and computers are locked (or even shut down) before employees leave, the cleaning crew might be allowed to enter the space.

87. Can we bring visitors into a room where CUI is accessible?

Yes, but only if the room has been properly "sanitized" of CUI (i.e., it has been put away and is not readily accessible to them) and they are escorted, or if those visitors:

- have a lawful government purpose to access that CUI,
- know how to handle CUI, and
- are not prohibited from receiving the CUI due to limited dissemination control.

88. How are the CUI program and the DoD CMMC program related?

The short answer is that the Cybersecurity Maturity Model Certification (CMMC) program involves the independent validation that a government contractor is ready to handle CUI, so that DoD can comply with 32 CFR 2002 and feel more confident when entrusting CUI to the contractor, or asking the contractor to create CUI.

89. Can you provide more details?

The more detailed version of the previous answer is that, as discussed in previous chapters, information is CUI in part because a law, regulation, or government-wide policy says that specific information must, or can, be safeguarded. The government already has its own, extensive information safeguarding requirements for protecting CUI and other sensitive information. The government's requirements involve performing complex risk analyses to determine appropriate safeguarding measures that must be put in place to protect the information handled by a federal system.

In its role as the Executive Agent for the CUI program, NARA recognized that most government contractors lacked the resources and training necessary to perform these risk analyses. In addition, since CUI is the government's information (or information entrusted to the government), the government is in the best position to perform the risk analyses.

So, NARA worked with the National Institute of Standards and Technology (NIST), as well as several federal agencies and industry partners, to perform the risk analyses and arrive at a standard set of safeguarding requirements that could be established for all non-federal IT systems that store, process, or transmit (i.e., *handle*) CUI. These non-federal IT systems include, for example, IT systems maintained by government contractors, state and local governments, and foreign governments. The result is NIST Special Publication (SP) 800-171.

The United States Department of Defense (DoD) has been at the forefront of the CUI program since (and even before) NARA published 32 CFR 2002. In October 2016, shortly after 32 CFR 2002 became effective, DoD published a new acquisition regulation, the Defense Federal Acquisition Regulation Supplement (DFARS) 252.204-7012 (DFARS-7012), which required all contractors who handle CUI on non-federal information technology (IT) systems to implement the requirements in NIST SP 800-171 on those IT systems.

DoD gave contractors over a year to implement the safeguarding requirements and comply with the new regulation. Unfortunately, during that time, and even a year or more afterward, DoD regularly found evidence that adversaries had been able to steal the government's sensitive information by breaching contractors' IT systems.

In investigating those breaches, DoD learned that, despite the contractual requirement, many of its contractors simply had not even begun to meet the requirements in NIST SP 800-171. This remained true long after the implementation deadline had passed. This left DoD in the difficult position of not meeting its own obligations under 32 CFR 2002. It also put the nation at risk, because adversaries were able to get access to critical information (e.g., designs for next-generation weapons, defensive equipment, munitions).

In 2019, DoD rethought its approach to ensuring that its contractors can properly handle CUI. DoD decided that allowing contractors to self-assess and attest that they comply with NIST SP 800-171 simply wasn't working. DoD thus created the CMMC program to help address the shortcomings in DFARS -7012.

Under the CMMC program, an independent, properly accredited third party must assess a given contractor's IT system, using the techniques identified in NIST's Assessment Guide for NIST SP 800-171, referred to as NIST SP 800-171A, as further refined by the CMMC Assessment Guide. If the independent assessor determines that the contractor complies with the requirements, the assessor's organization, referred to as a Certified 3rd Party Assessment Organization (C3PAO), issues a certification to the contractor and sends a copy to DoD.

As part of their procurement process, when the CMMC program is fully in effect, DoD Contracting Officers will be required to ensure that certifications are in place for all contractors handling CUI under a given proposal *before* DoD awards the prime contractor a contract for that proposal. This gives DoD higher levels of assurance that the nation's information is better safeguarded when it is in the hands of DoD contractors and other third parties.

Chapter 11: Government Contractors and CUI

90. How should contractors handle legacy information?

Legacy information must be handled in accordance with the terms of the contract under which it was created or received. If the contractor creates a derivative work based on legacy information, the contactor should contact their prime contractor or the point of contact identified in their contract or SCG to determine if the derivative work contains CUI.

If the contractor subsequently is told that certain legacy information is now CUI, then *all* copies of that information must be marked and handled as CUI. (See, e.g., 32 CFR 2002.20 and NARA CUI FAQ "When it comes to legacy information, should a contractor/company," etc.)

91. Are government contractors authorized to designate information as CUI?

In general, no. There may be exceptions, but in most cases only properly authorized agency personnel can designate information as CUI. Even when contractors are creating information for the government, the *agency* must designate that information as CUI. The agency must also provide the CUI designation block and CUI banner marking(s) as part of this process. (See, e.g., 32 CFR 2002.20(d)-(f).)

92. How do agencies tell contractors that information the contractor will create has been designated as CUI?

Agencies typically use contracts with their contractors to designate as CUI any information the contractor will create under that contract. Some agencies, such as DoD, require that CUI designations for information a contractor is expected to create must be communicated as part of a

Security Classification Guide. This Guide is incorporated into the contract with the contractor. (See, e.g., DoDI 5200.48 Section 3.7.e.)

Other agencies may take a less structured approach. For example, if the contractor enters into a contract to create a new widget for the Department of Good Works (DGW) based on a DGW requirements specification, DGW may instruct the contractor that:

> design specifications, drawings, source code, testing procedures, testing results, and related technical documents which define or relate to the widget(s) created under this contract have been designated as CUI//SP-CTI. All copies of such information, including any media containing the information, must include banner markings consistent with the CUI marking guidance provided by the National Archives and Records Administration. Wherever feasible, all copies of the information must also include a designation block which incorporates the following information on at least the cover page or exterior of any related media: Designating Agency: Department of Good Works; Division; ....

As an aside, when agencies communicate CUI designation information to contractors, they should ensure that such communications are provided in *unclassified* documents. In such scenarios, contractor personnel without security clearances or need to know cannot review CUI designation information in those documents. As a result, contractors may inadvertently not mark information they create as CUI because the appropriate personnel were not aware of the CUI designation. NARA's CUI Marking Guide includes guidance on how to handle scenarios in which CUI and classified information must coexist in the same document (e.g., by creating supplements or appendixes with the classified information that can be removed when the Security Classification Guide must be shared with non-cleared individuals).

93. Are government contractors authorized to mark information as CUI?

Yes. All government contractors are authorized to mark information they create as CUI if that information has been designated as CUI. It must be marked using the markings specified by the agency for which the contractor is working, or by the designating agency. (See, e.g., 32 CFR 2002.20(d) and NARA CUI FAQ, "Does Industry ever have to mark CUI?")

In addition, if the contractor is instructed by the government that certain (e.g., legacy) information has been designated as CUI, the contractor should ensure it is properly marked as CUI, and that any legacy markings are removed from that information. (See, e.g., 32 CFR 2002.36.)

94. Are government contractors authorized to designate their own information as CUI?

No. Regulation 32 CFR 2002 makes clear that only *agencies* of the executive branch of the federal government can designate information as CUI. Since a government contractor is *not* an agency, they cannot designate their information as CUI. (See, e.g., 32 CFR 2002.04(a).)

Contractors can, however, mark their information as "Proprietary Information" and include a cover letter explaining that the agency to which it is being submitted *should* designate and mark the information as CUI under one or more LRGWPs.

95. Can a government contractor's information be CUI?

Yes, in some circumstances, but not while it is in that contractor's systems. For example, a contractor may design, at its own expense and of its own initiative, a new tool that it plans to sell. The design for that tool is proprietary to the contractor and is not CUI, because it was not created for or on behalf of the government (i.e., by definition it cannot be CUI).

If the government wants to buy some tools made according to that design, the government might ask for a copy of the design to evaluate the suitability of the tool. In that case, the government might designate and mark the design document as CUI (e.g., under 18 USC 1905 or 18 USC 1832) when it is in the government's systems.

96. If a government contractor receives their own information from the government, is it CUI?

No. Extending the example in the previous question, even if the government sent a copy of the tool design document back to the contractor with the design document marked as CUI, since the information is the contractor's own information, the design document is not CUI while in the contractor's system. (See, e.g., 32 CFR 2002.4(h).)

97. If a government contractor receives information about another contractor from the government, can it be CUI?

Yes. If the government has designated and marked information as CUI, it is CUI. The only exception is when the government returns the contractor's own information back to the contractor.

Continuing the scenario from the previous question, if the government hired another contractor to analyze the suitability of the design, the government should designate and mark the design document as CUI. The other contractor would then be required to treat that information as CUI.

98. How should contractors handle suspected CUI?

If you receive information that you suspect is CUI, but it isn't marked as CUI, you should perform both these steps:

1. Treat (but do *not* mark) it as though it is CUI until you are told otherwise. As discussed in question 51, you may want to add a coversheet to the information that indicates that it is "Suspected

CUI—Designation Pending," but you should not mark the information without further authorization.

2. Contact the disseminating agency (i.e., the agency that gave you the information, typically through the Contracting Officer or the agency's CUI program management office) to confirm the information is not CUI.

(See, e.g., 32 CFR 2002.20(m) and 32 CFR 2002.50.)

99. Should contractor proposals include any CUI-related provisions?

Yes, probably. Although the CUI program has been in existence since 2008, many agencies are just beginning to implement its requirements. It could be advantageous for contractors to include language in the Assumptions section of their proposals along the following lines:

Contractor recognizes that any contract resulting from this proposal is likely to involve the creation or handling of CUI by contractor and/or a subcontractor. Consistent with the agency's obligations under 32 CFR 2002, contractor assumes that the agency will properly mark as CUI all information which the agency has designated as CUI prior to its dissemination to contractor.

In addition, contractor assumes that, consistent with the agency's obligations under 32 CFR 2002, the agency has previously communicated, and will continue to communicate, to contractor regarding any information that contractor may create under this contract which the agency has designated as CUI.

Contractor further assumes that the agency recognizes that the agency's failure to properly designate information as CUI, mark information as CUI, and/or communicate appropriate designation and marking information to contractor in a timely manner may result in schedule delays, may increase contractor's costs of performance, and may have other impacts on the overall performance of the contract which will be addressed by the agency in the form of one or more contract modifications.

Because of the variety of issues that could arise, contractors should consult with a government contracts lawyer or other professional regarding the appropriateness of such language in their proposal.

100. Are there any other CUI-related concerns for government contractors?

Yes, there are many! For example, DoD contractors who use cloud services to handle CUI must ensure that those cloud services are authorized to operate at the Federal Risk and Authorization Management Program (FedRAMP) moderate level (or have implemented security controls equivalent to this level). In addition, DoD contractors must ensure that those cloud services agree to notify the contractor of any incidents or suspected breaches within a specific timeline, and allow DoD to audit/inspect the service provider's data center(s) where the CUI is stored when there is a breach or other incident. (See, e.g., DFARS 252.204-7012.)

DoD contractors who use a managed IT or security service provider should be aware that their service provider's systems may also be subject to review and approval in certain circumstances. When the CMMC program comes into effect, that program will also require independent third-party certification that the service provider's system meets the requirements in NIST SP 800-171 in certain cases. (See, e.g., DFARS 252.204-7021 and the CMMC Level 2 Scoping Guide.)

References

Acronyms

Acronym	Meaning
C3PAO	Certified 3rd Party Assessment Organization
CDI	Covered Defense Information
CFR	Code of Federal Regulations
CUI	Controlled Unclassified Information
DFARS	Defense Federal Acquisition Regulation Supplement
DoD	Department of Defense
FAR	Federal Acquisition Regulations
FCI	Federal Contract Information
FedRAMP	Federal Risk and Authorization Management Platform
FIPS	Federal Information Processing Standard
FOUO	For Official Use Only
ISOO	Information Security Oversight Office
LES	Law Enforcement Sensitive
LRGWP	Law, Regulation, or Government-Wide Policy
NARA	National Archives and Records Administration
NIST	National Institute of Standards and Technology
RFP	Request for Proposals
SBU	Sensitive but Unclassified
SP	Special Publication
USC	United States Code

Sources by question

#	Text	Link
6	NARA's Training	https://www.archives.gov/cui/training.html#intro-to-marking
6	DoD's CUI Marking Training	https://securityawareness.usalearning.gov/cui/index.html
8	9/11 Commission	https://www.9-11commission.gov/report/911Report.pdf
11	Executive Order 13556	https://obamaWhite House.archives.gov/the-press-office/2010/11/04/executive-order-13556-controlled-unclassified-information
12	publicly complained	https://irp.fas.org/offdocs/nsm/nsc-review.pdf
12	National Cybersecurity Strategy	https://www.White House.gov/wp-content/uploads/2023/03/National-Cybersecurity-Strategy-2023.pdf
12	Executive Order (EO) 14028	https://www.White House.gov/briefing-room/presidential-actions/2021/05/12/executive-order-on-improving-the-nations-cybersecurity/
17	32 CFR 2002	https://www.ecfr.gov/current/title-32/subtitle-B/chapter-XX/part-2002
20	Federal Acquisition Regulation (FAR) 52.204-21(a)	https://www.acquisition.gov/far/52.204-21
25	CUI Registry	https://www.archives.gov/cui/registry/category-list
25	DoD's CUI Registry	https://www.dodcui.mil/CUI-Registry-New/
28	Defense Federal Acquisition Regulations Supplement (DFARS) 252.204-7012	https://www.acquisition.gov/dfars/252.204-7012-safeguarding-covered-defense-information-and-cyber-incident-reporting.
29	NIST Special Publication 800-171	https://csrc.nist.gov/pubs/sp/800/171/r2/upd1/final
34	DoD Instruction 5230.24.2.8.e	https://www.esd.whs.mil/portals/54/documents/dd/issuances/dodi/523024p.pdf
34	DoD Instruction 5200.48.3.8.d	https://www.esd.whs.mil/Portals/54/Documents/DD/issuances/dodi/520048p.PDF
42	CUI Marking Handbook	https://www.archives.gov/files/cui/documents/20161206-cui-marking-handbook-v1-1-20190524.pdf

45	free CUI-related training videos	https://www.archives.gov/cui/training.html#intro-to-marking
45	CUI marking training	https://securityawareness.usalearning.gov/cui/index.html
67	CMMCInfo.org	Https://CMMCInfo.org
68	CUI Registry (limited dissemination controls)	https://www.archives.gov/cui/registry/limited-dissemination
71	CUI points of contact	https://www.archives.gov/cui/about/contact.html#contact-an-agency
80	NIST SP 800-171A	https://csrc.nist.gov/pubs/sp/800/171/a/final
89	Cybersecurity Maturity Model Certification (CMMC)	https://dodcio.defense.gov/CMMC/Documentation/
90	NARA CUI FAQ	https://www.archives.gov/cui/faqs.html
100	CMMC Level 2 Scoping Guide	https://dodcio.defense.gov/CMMC/Documentation/

About the Author

James ("Jim") Goepel is a lawyer (JD, LLM, The George Mason University School of Law) and engineer (BSECE, Drexel University). He earned Provisional Instructor, Provisional Assessor, Certified CMMC Professional, and Certified CMMC Assessor certifications from the CMMC Accreditation Body (the "Cyber AB").

Jim has had a diverse career in and out of government. Prior to law school, Jim was a civilian engineer with the United States Coast Guard and worked in the civilian space program. He was also a software developer on US government contracts, and developer and systems administrator for several organizations, including for the US Congress, where he managed and secured systems which processed approximately 85% of the legislation that came before Congress.

During and since law school, Jim has advised government contractors, from small businesses to $1B+ organizations, on legal, cybersecurity, business operations, IT, intellectual property, technology transfer, and other topics.

Jim is a cofounder of the CMMC Information Institute, a non-profit (501(c)(3)) educational organization. He is also a former professor at Drexel University, where he created and taught undergraduate and graduate cybersecurity courses.

Jim is also one of the Founding Directors of the Cyber AB. During his tenure, he was elected as Board Treasurer and created and taught the ecosystem's first official CMMC educational program.

Acknowledgments

I have a *lot* of people to thank for their support and encouragement, including:

- My wife, Beth Montagna, for her invaluable edits and feedback on early drafts of the book, as well as her patience reading through the same material multiple times! I think she knows the CUI program better than me at this point!
- My family, for their patience as I disappeared into my office for many nights and weekends while researching and writing both books.
- Fernando Machado, of Cybersec Investments, for encouraging me to stick with it and write this book after completing *CUI Informed*, as well as his feedback on the structure and content of both books.
- Ret. Col. Greg Gardner, for allowing me to interview him about the early days of the CUI program. That interview helped confirm some things I had inferred and helped me better understand several others about the purpose behind the CUI program.
- Charlene Wallace for her amazing insights into the CUI program and the content of *CUI Informed*, which also found its way into this book.
- Michael Connelly for his invaluable feedback on this book.
- Kyle Lai, of KLC Consulting, for his feedback on this book, and especially the government contractor chapter.
- My amazing colleagues and friends across the cybersecurity, compliance, CMMC, and academic communities, including Tara Lemieux, Matthew Titcombe, Ben Tchoubineh, Amy Starzynski-Coddens, Mark Berman, Justin Pelletier, Kenny Houston, Regan Edens, Chase Berman, Jordan Fischer, and Paul Flanagan. Without their support, encouragement, feedback, and in some cases our occasionally spirited discussions, *CMMC Fundamentals* and *CMMC Informed* would not have happened!

www.ingramcontent.com/pod-product-compliance
Lightning Source LLC
Chambersburg PA
CBHW040909210326
41597CB00029B/5021